Natural Disasters

TORNADOES

CHRISTINE THOMAS ALDERMAN

BLACK RABBIT BOOKS

Bolt is published by Black Rabbit Books
P.O. Box 3263, Mankato, Minnesota, 56002.
www.blackrabbitbooks.com

Marysa Storm, editor; Catherine Cates, designer;
Omay Ayres, photo researcher

Library of Congress Cataloging-in-Publication Data
Names: Alderman, Christine Thomas, author.
Title: Tornadoes / by Christine Thomas Alderman.
Description: Mankato, Minnesota : Black Rabbit Books, [2021] | Series: Bolt. natural disasters | Includes bibliographical references and index. | Audience: Ages 8-12 | Audience: Grades 4-6 | Summary: "Take a close look at tornadoes through engaging text, fun facts, and infographics that stress visual literacy"- Provided by publisher.
Identifiers: LCCN 2019026703 (print) | LCCN 2019026704 (ebook) | ISBN 9781623102623 (hardcover) | ISBN 9781644663585 (paperback) | ISBN 9781623103569 (ebook)
Subjects: LCSH: Tornadoes-Juvenile literature.
Classification: LCC QC955.2 .A425 2021 (print) | LCC QC955.2 (ebook) | DDC 551.55/3-dc23
LC record available at https://lccn.loc.gov/2019026703
LC ebook record available at https://lccn.loc.gov/2019026704

Printed in the United States. 2/20

Image Credits

Alamy: FEMA, 24–25; Ryan McGinnis, 21; UPI, 17; en.wikipedia.org: Red Thunder, 14; iStock: aapsky, 22; PeteDraper, 6; Newscom: Fema, 18–19; Greg Johnson / Mediadrumworld.Co, 4–5; NOAA: Natnional Weather Service, 22–23; Science Source: Gary Hincks, 10, 10–11, 11, Jim Reed, 28; Shutterstock: aphichato, 31; Cammie Czuchnicki, 9; Fer Gregory, 13; John D Sirlin, 3, 21; Kattiya.L, 26; Mette Fairgrieve, 22; Noel V. Baebler, 27; ohenze, 32; photka, 26; Rainer Lesniewski, 12–13; solarseven, Cover, 1; Vereshchagin Dmitry, 22–23; sosyalforum.org: CASPER CARTER, 27

Every effort has been made to contact copyright holders for material reproduced in this book. Any omissions will be rectified in subsequent printings if notice is given to the publisher.

CONTENTS

CHAPTER 1

A Huge STORM

Sirens wail. The sky grows dark. People rush to storm shelters and basements. A funnel cloud spins from the clouds to the ground. Tornado! The air roars like a train. **Hail** pounds the ground. Minutes seem like hours. Once the storm is done, the sky clears. The tornado is gone. But a path of destruction remains.

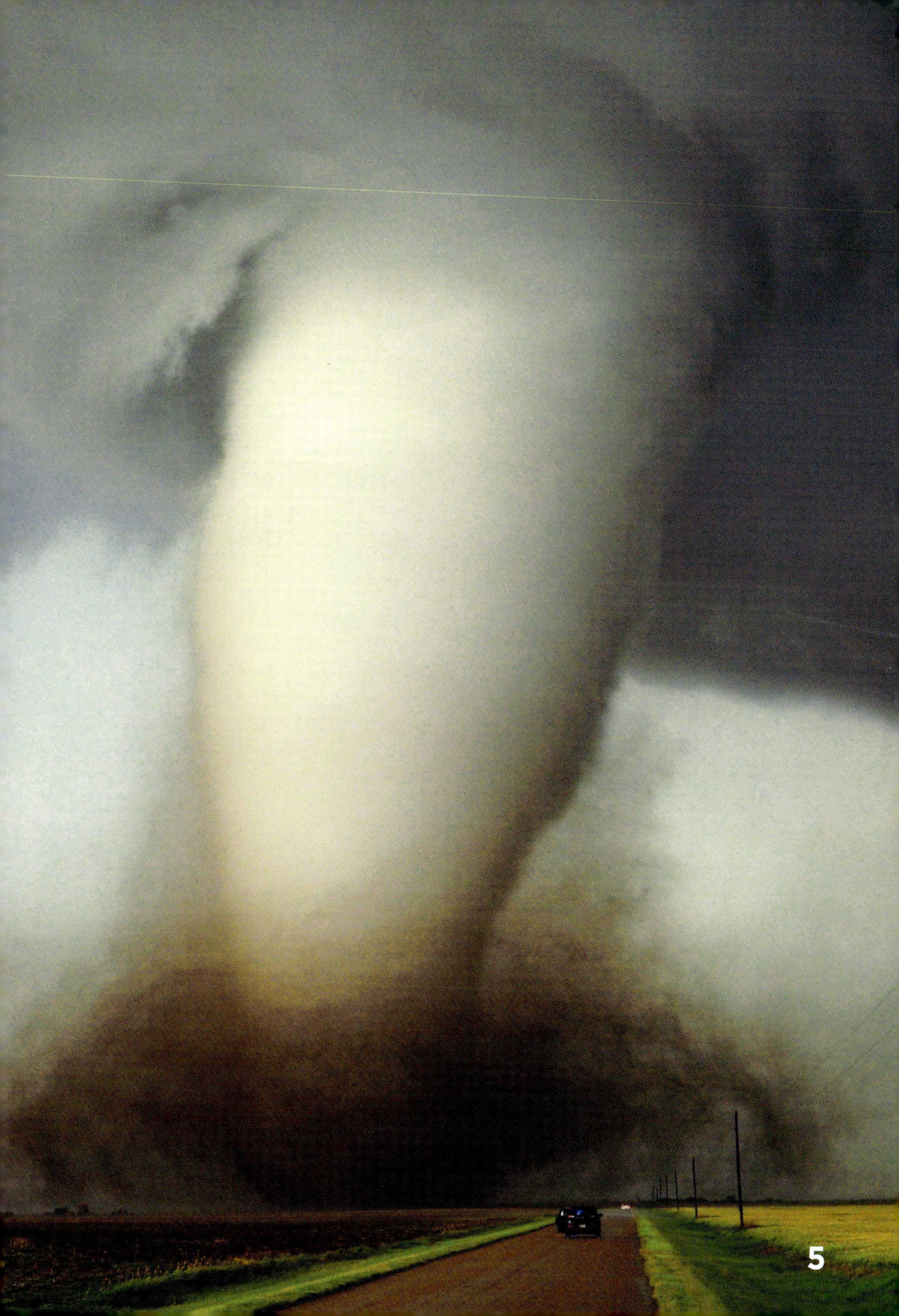

Tornadoes can form over water. They're called waterspouts.

What Is a Tornado?

Tornadoes, also called twisters, are deadly columns of spinning air. They're hundreds of feet wide. Their winds can race more than 300 miles (483 kilometers) per hour. These winds destroy houses. They turn **debris** into fast-flying **missiles**.

Most tornadoes are fairly weak. They're on the ground for only a few minutes. They move at about 15 miles (24 km) per hour. They travel less than 6 miles (10 km). But some strong storms produce giant, fast tornadoes. These tornadoes can last hours and move at 70 miles (113 km) per hour.

CHAPTER 2

What Causes TORNADOES?

Tornadoes often form in spring and summer. They start in thunderstorms. When storm clouds rotate, they can create funnel clouds. **Pressure** inside the swirling cloud columns then pushes them to the ground. Tornadoes can form quickly. Sometimes, it only takes seconds.

The swirling winds in a hurricane can start tornadoes.

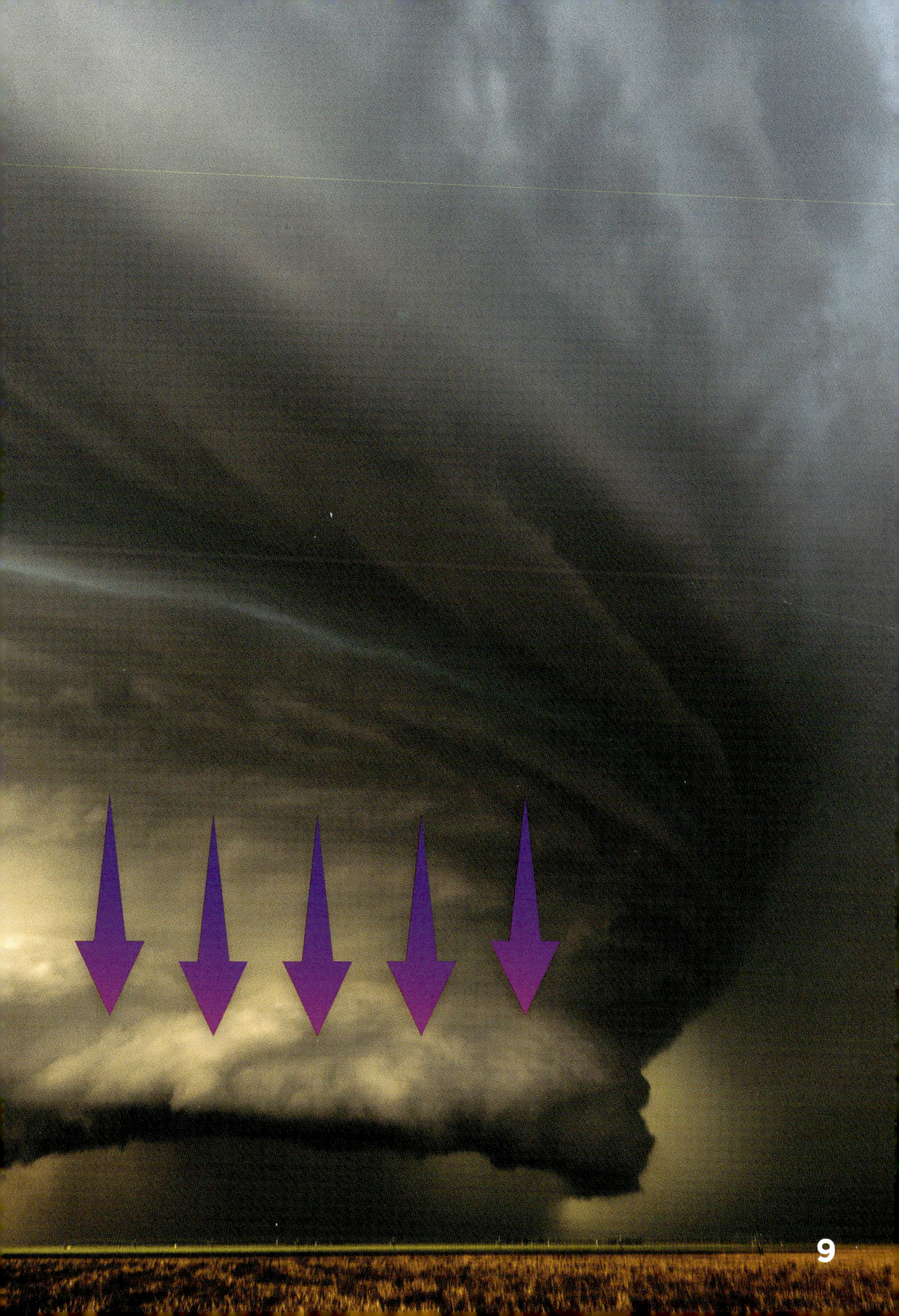

Wind going in different directions causes air within a storm to spin horizontally.

Step 1

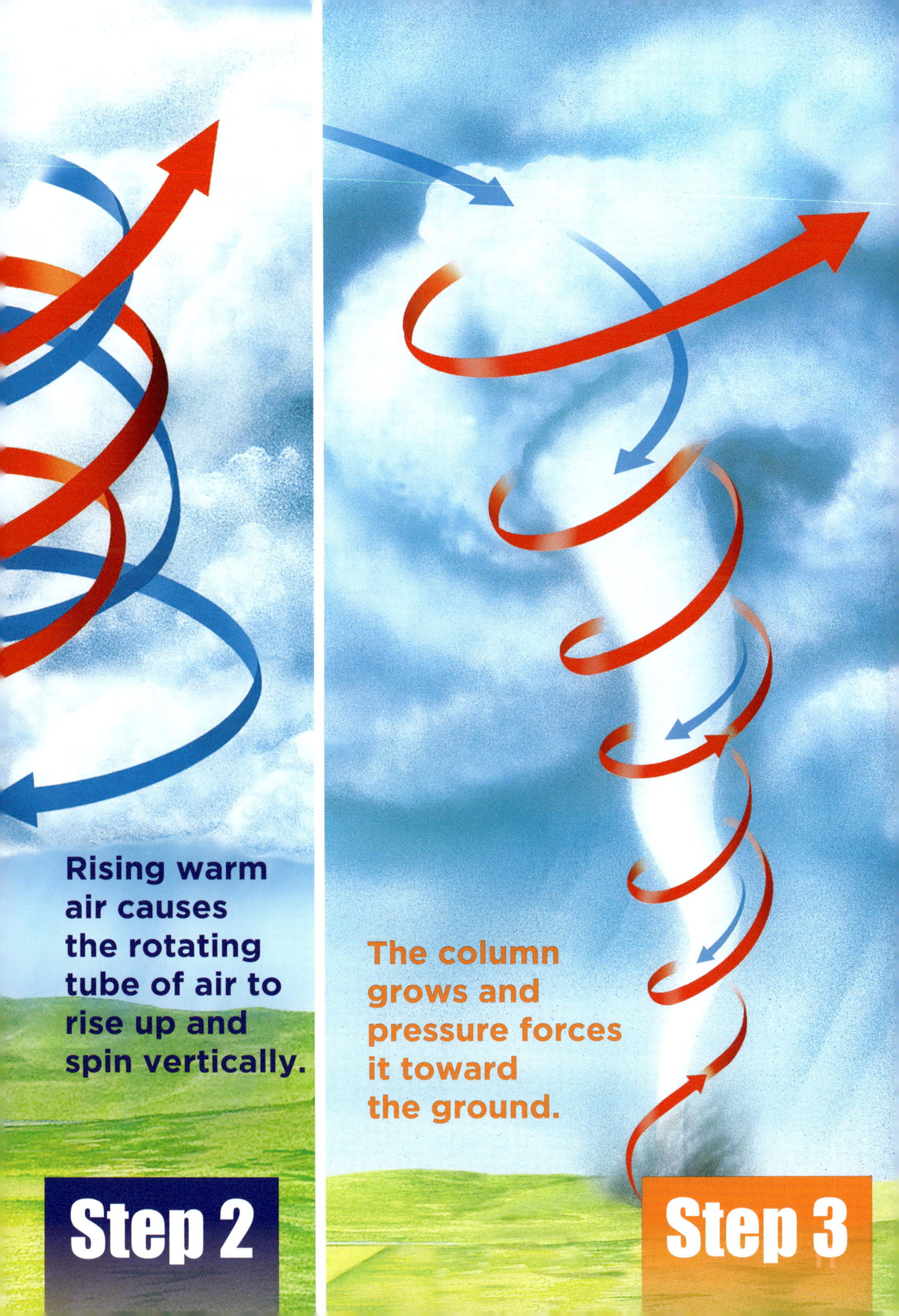
Rising warm air causes the rotating tube of air to rise up and spin vertically.
Step 2
The column grows and pressure forces it toward the ground.
Step 3

Where They Happen

Tornadoes strike worldwide. But about 75 percent of tornadoes happen in the United States. Cold polar air and warm tropical air often meet there. This meeting creates storms. The country has more than 1,000 tornadoes yearly. Most twist across Texas. Tornadoes often hit the United Kingdom too. The small country has the most tornadoes per land size. They're usually weak.

Tornado Alley

Tornado Alley is an area in the United States. This area got its nickname because many twisters form there.

CHAPTER 3

Famous TORNADOES

Tornadoes cause extreme damage. In 1989, the deadliest tornado ever recorded struck Bangladesh. Its path was 10 miles (16 km) long and 1 mile (2 km) wide. The tornado destroyed more than 20 villages. It left about 80,000 people homeless. About 1,300 people lost their lives.

The Joplin Twister

On May 22, 2011, a twister ripped through Joplin, Missouri. It was an EF-5. The tornado killed more than 150 people and hurt 1,000. Thousands of homes were torn apart. The storm cost the city about $3 billion.

ENHANCED FUJITA SCALE

Scientists use this scale to rank tornadoes based on wind speed.

EF-0

65 to 85 miles (105 to 137 km) per hour

EF-1

86 to 110 miles (138 to 177 km) per hour

EF-2

111 to 135 miles (179 to 217 km) per hour

EF-3

136 to 165 miles (219 to 266 km) per hour

EF-4

166 to 200 miles (267 to 322 km) per hour

EF-5

greater than 200 miles (322 km) per hour

Two schools were badly damaged. Ten children died.

The Moore Tornado

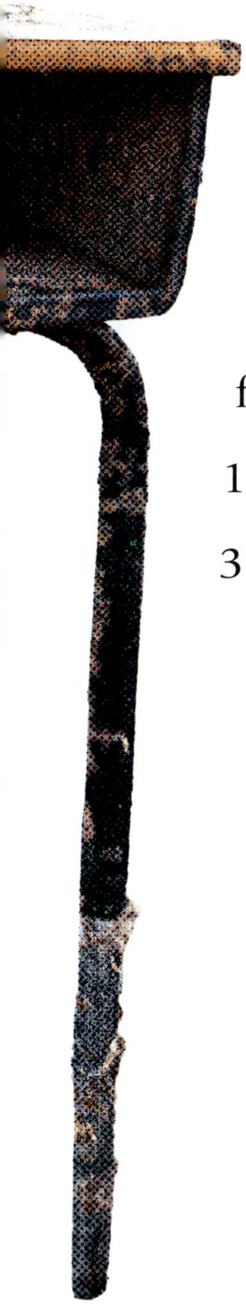

A tornado struck Moore, Oklahoma, on May 20, 2013. It was also an EF-5. Wind speeds topped 210 miles (338 km) per hour. It **churned** across the ground for about 40 minutes. The path was 17 miles (27 km) long. It injured 377 people and killed 24.

CHAPTER 4

Preparing for TORNADOES

Tornadoes are deadly. But spotting storms saves lives. Scientists watch the weather. They look for weather patterns that might lead to tornadoes. When it looks like tornadoes might form, they issue watches. When tornadoes form, they send out warnings.

WATCHING THE WEATHER

Scientists use these tools to watch growing storms.

COMPUTER
MODELING
DOPPLER
RADAR

Taking Shelter

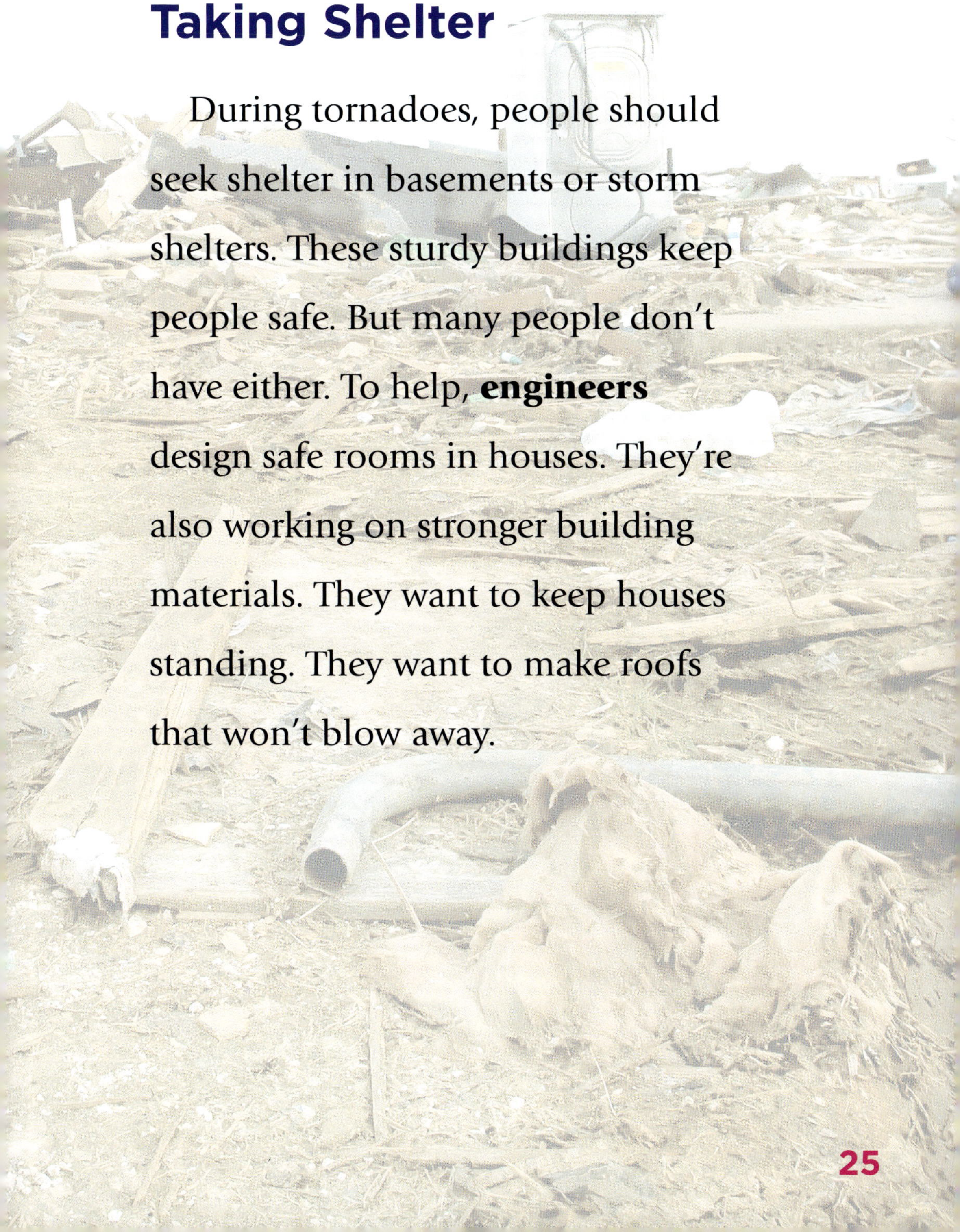

During tornadoes, people should seek shelter in basements or storm shelters. These sturdy buildings keep people safe. But many people don't have either. To help, **engineers** design safe rooms in houses. They're also working on stronger building materials. They want to keep houses standing. They want to make roofs that won't blow away.

STAYING SAFE DURING A TORNADO

Have an emergency kit and disaster plan ready.

Listen to weather reports and sirens.

Go into a basement, storm shelter, or room with thick walls and no windows.

Duck and cover, using arms to protect head and neck.

Powerful Storms

Tornadoes are deadly disasters. They take lives and destroy homes. But being prepared prevents death and damage. To help, people work to make stronger buildings. They keep studying the weather. Faster warnings will give people more time to get ready. Fewer lives will be lost. These strong storms affect people around the world. Their power is both awesome and awful.

churn (CHURN)—to move by or as if by forceful stirring action

debris (duh-BREE)—the pieces that are left after something has been destroyed

Doppler radar (DOP-ler RAY-dar)—a device that is used for predicting weather

engineer (en-juh-NEER)—a designer or builder

hail (HEYL)—small lumps of ice and snow that fall from clouds sometimes during thunderstorms

missile (MIS-uhl)—a weapon that can be thrown or projected to hit an object far away

pressure (PRESH-er)—the weight of the air in the earth's atmosphere

satellite (SA-tuh-liyt)—a machine sent into space that moves around the earth, moon, sun, or other planet

BOOKS

Levy, Janey. *Devastating Storms*. Mother Nature Is Trying to Kill Me! New York: Gareth Stevens Publishing, 2020.

Lusted, Marcia Amidon. *The 12 Worst Tornadoes of All Time*. All-Time Worst Disasters. Mankato, MN: 12-Story Library, 2019.

Rathburn, Betsy. *Tornadoes*. Natural Disasters. Minneapolis: Bellwether Media, Inc., 2020.

WEBSITES

Earth Science for Kids: Weather – Tornadoes
www.ducksters.com/science/earth_science/tornadoes.php

Fun Tornado Facts for Kids
www.sciencekids.co.nz/sciencefacts/weather/tornado.html

Tornado
kids.nationalgeographic.com/explore/science/tornado/#tornado.jpg

INDEX